HAND-BOOK

FOR THE

WAR.

DESCRIBING THE MILITARY TERMS IN USE IN THE UNITED STATES SERVICE, AND GIVING A LIST OF THE FORTS AND SHIPS BELONGING TO THE UNITED STATES, TOGETHER WITH A PARTICULAR DESCRIPTION OF THE MORE IMPORTANT ONES.

BOSTON:
WHOLESALE OFFICE, No. 21 CORNHILL.
1861.

Press of Geo. C. Rand & Avery, No. 3 Cornhill.

For the information obtained by me, while writing this pamphlet, I am greatly indebted to several standard authors, and two or three army officers. It has been my intention to give an idea of the meaning of most of the words used in our Service, and coming to us through the newspaper reports of the times.

I have given merely the military definition of those words used with both a military and civil signification, and have omitted those nearly obsolete, or in use in foreign countries.

GIHON.

HAND-BOOK FOR THE WAR.

MILITARY TERMS AND THEIR DEFINITIONS.

ADJUTANT — An officer whose duty it is to assist higher officers in receiving and communicating orders. He places guards, distributes ammunition, etc.

ARMORY — A place used for the manufacture of weapons.

ARSENAL — A place used for the storage of weapons.

AMMUNITION — Articles used in the discharge of fire-arms.

AIR-GUN — A gun made with a hollow stock, which is used as a receiver for compressed air, which is let into the barrel by a cock which forces out the ball, — it having been loaded with a wad in the ordinary way. In an improved gun, gas is ignited by an electric spark, which expels the ball with about the force of powder.

ARTILLERY — The name of a body of troops using cannon for weapons.

BARRACK — A building for the lodgement of soldiers, together with kitchens, mess-room, hospital, and in some cases, library and reading-room. If cavalry are

1*

to be accommodated, the stables for the horses are connected.

Battery — In field artillery, the term is used to designate any number of guns from four to twelve (with their necessary equipments), which are intended to act together in battle.

Field batteries are divided into light, heavy, and howitzer batteries. The word "battery" is also used to designate the spot where guns are placed. A battery is either elevated, sunken, or half-sunken, and composed of either guns, howitzers, or mortars.

To construct an earthwork battery, the principal dimensions are traced and the earth procured from a ditch in front or rear of the intended parapet. The inside is built up steep by the use of sods of turf, sand-bags, casks of earth, etc., while the outside is left sloping to give additional strength to the wall, which sometimes is twenty-four feet thick.

The ammunition is kept partly in recesses under the wall, and partly in a sunken building of timber, covered in, bomb-proof, with earth.

Floating batteries are sometimes made of old men-of-war, with their sides strengthened to resist shot, and sometimes they are mere rafts with walls erected.

A formidable battery is being constructed at Hoboken, N. J. Her principal dimensions are, length, 415 feet; breadth, 48 feet; depth, 32 feet, 4 inches; steam-power, 8,624 horses. The vessel is constructed entirely of iron, her ends are very sharp, and her speed, when steaming at full power, will be very great. Neither the armament nor the exact thickness of the protected portions of the vessel is fully determined. The defences will probably be composed of very thick forged-iron plates, or thinner plates riveted together.

The battery is intended for New York Bay, from Sandy Hook upward, and is now nearly completed. All her machinery and boilers are in place.

Battalion — In America, a battalion is composed of eight companies of infantry of the line.

Bayonet — A pointed weapon, fastened on a gun near the muzzle.

Bivouac — An encampment of troops by night in the open air, without tents, each soldier ready armed and sleeping in his clothes with, but sometimes without, straw.

Bastion — A tower of a fort with generally more than four faces.

Bomb, or Shell — A hollow shot, filled with powder, and sometimes pieces of iron or other hard substances. They are intended to explode when they reach the object at which they are fired, and are very effective when fired so as to fall from the greatest possible height.

They are usually made from ten to twenty-four inches in diameter, and when the smaller sizes are charged with from six to eight pounds of powder, the fragments are forced nearly three thousand feet if unobstructed, as when bursting near the surface of a hard material; but when the bomb falls into soft soil, it often imbeds itself ten or twelve feet deep, so that the explosion is comparatively harmless.

Bomb Ketch — An old-fashioned war-vessel for carrying mortars.

Bomb-proof — A term used to designate a roof strong enough to resist the action of bombs falling upon it. Such roofs are generally used to cover in magazines and hospitals, and are covered with iron masonry, and sometimes several feet of earth.

Bomb Vessel — A vessel constructed to carry mortars with as light a draft of water as possible.

Bombardment — The act of throwing bombs into a fortress or town. This method of warfare is found very effective in destroying the internal arrangements of a fortress, or buildings in a city.

Bombast — A use of lofty words, most indulged in by those who would be quickest to dodge a bomb.

Bowie Knife — A long single-edged knife.

Breastwork — An elevation (usually on a hill) for the protection of troops against an enemy; generally a mass of earth, but sometimes built of bags of sand, cotton, wool, etc., and made upwards of ten feet thick.

Brevet — In the United States, a term used implying a higher nominal rank than that for which pay is received; thus, a brevet-major receives captain's pay.

Brig — A vessel with two square-rigged masts.

Brigade — In the United States army, two regiments of infantry or cavalry, commanded by a brigadier-general.

Banquette — Steps for infantry to stand upon when firing over a fort wall from within.

Company — In infantry, sixty-four men, with their officers.

Camp — A place of repose for troops for one night or longer, with or without shelter.

Cannon — An implement of war for throwing heavy projectiles, as shot and shells.

Dahlgreen's 8-inch shell-gun, — length of bore, 100.3-inch; weight, 63 cwt.

Cutlass — A short, heavy sword, mostly used in hand conflicts, by marines.

Cannonade — An engagement between two armies, in which the artillery alone is active.

Captain — A commander of a company of infantry, or of a troop of cavalry, or the chief officer of a ship of war.

Cartel — A writing or agreement between hostile persons for some mutual advantage, such as the exchange of prisoners.

Cartel Ship — A vessel used in exchanging prisoners, or carrying proposals to an enemy.

Cartouch — Cartridge-box.

Curtain — Piece of wall between two towers of a fort.

Case Shot — Musket-balls or other hard substances, put in cans to be fired from cannon.

Cartridge — A case or bag containing the exact quantity of gunpowder used for the charge of a fire-arm. A blank cartridge does not contain a bullet.

Cavalry — A body of soldiers on horseback in the United States service. They are really mounted infantry.

Chain Shot — Two balls connected by a chain, mostly used in naval battles, — used to cut down masts and rigging.

Chaplain — A person appointed by Government as minister to a body of soldiers.

Charge — A rapid advance of soldiers on the enemy, with the intention of breaking their ranks by the momentum of the attacking party.

Counter-mure — A wall raised behind another.

Counter-march — To march back.

Colonel — The chief commander of a regiment, in rank between a brigadier-general and lieutenant-colonel.

Column — A body of troops arranged with a narrow front and deep files.

Commissary — An officer whose duty it is to provide food and conveyance for an army.

Commission — A writing by which an officer holds his appointment.

Cornet — An officer of cavalry bearing the colors of the troops.

Corporal — The lowest officer in a company of infantry, between a private and a sergeant. He does duty in the ranks as private, but has charge of a squad at drill.

Draft — Drawing of men.

Dudgeon — A small dagger.

Dragoons — A company of one kind of cavalry.

Dahlgreen Gun — A cannon of peculiar pattern, for firing shell. (*See* Cannon.)

Drawn Battle — One without victory on either side.

Enfilade — To scour with shot the length of a line. A battery is sometimes arranged to fire its guns so that the balls will pass just inside the enemy's fortifications, and parallel to its walls, endangering men and guns the length of the line.

Embargo — A public law forbidding ships to sail.

Escalade — An attack on a fortified place by mounting the walls with ladders without the formalities of a siege.

Escarpment — The outside slope of a fortress wall.

Encampment — The regular arrangement of tents, or temporary shelter for the part or whole of an army.

Frigate — The name of a war-vessel, generally two-decked, and carrying fifteen guns and upward. Lately, they are propelled by steam at a high speed.

Fortification — A general term applied to structures for the protection of a point against an enemy, or for the reduction of other fortifications.

Gangrene (Hospital) — A putrid disease caused by crowding sick or wounded men into close rooms.

Globe — A body of soldiers formed into a circle.

Grape Shot — The name of a grouping of shot around a metallic spindle, the whole arranged to fit the bore of the cannon.

The shots fly asunder as they leave the gun, and are most destructive at short distances.

Grenade — A small iron shell, about two and a half inches in diameter, filled with combustibles. It is thrown with the hand, and is most used in close naval actions.

Guard — The designation of a body of troops detailed to watch and defend the post, or the main body of troops, from surprise, and to put down disorders. On the march, the vanguard precedes, and the rear-guard follows the main army.

Garrison — Fortification with its troops, or the troops belonging to a fortified place.

Glacis — Sloping bank of earth towards the field.

Gun — The term in common use is applied to every kind of a fire-arm except rifles and pistols.

The manufacture of guns for infantry has been most systematized at the government armories in this country. Each piece being made after a model, so that the smallest screw used in making one gun is precisely like, in shape and size, the screw used in the same part of either gun. The capacity of the Springfield and Harper's Ferry armories are twenty thousand muskets each per annum.

Glure — A broadsword.

GUNBOAT — In the United States Navy they are built of one thousand tons burden and one thousand horse-power. They are intended to carry six heavy guns.

GUNPOWDER — An explosive compound of sulphur, nitre, and charcoal.

GENERAL — Chief commander of an army.

HORSE PISTOL — Large pistol carried by cavalry.

HOWITZER — A kind of cannon used in firing shell. It is mounted on wheels, and used in field service.

INTRENCHMENT — As generally used, means a ditch, with its mound of earth.

INFANTRY — The foot soldiers of an army.

LANCERS — Horsemen armed with spears, or swords and pistols.

MAJOR — An officer next above a captain.

MINNIE RIFLE BULLET — This bullet was invented by a French officer, and consists of a cylindrical shot, conical in front and hollow behind, and fitted with a cap of thin iron which fills the grooves of the rifle as it is forced through, and thus gives great precision of flight to the ball. This bullet, with its rifle, is in very general use in Europe.

MILITIA — Able-bodied men between eighteen and forty-five years of age.

MORTAR — A short cannon of large bore, intended

for throwing shells. Those of from 10 to 13 inch bore are most used. They are usually fired at an elevation of about 45°, so that it is not generally necessary to have port-holes in the walls, from behind which they are fired. Very large sizes that have been introduced into different services have not worked satisfactorily.

Marine — A naval soldier.

Musket — A small bore gun of shorter range than a rifle.

Mine — A pit under a fortification intended to receive powder for blowing it up.

Muster — An assembling of troops for review.

Navy — All the ships of war belonging to a nation.

Private — A common soldier.

Port-Hole — Opening in the side of a ship through which cannon are discharged.

Platoon — A line of soldiers.

Privateer — A vessel of war commissioned by Government and owned by private individuals.

Port Fire — A composition for setting fire to powder, used instead of a match.

Parapet — A wall of earth or other material used to protect soldiers from an enemy's fire.

Parade — To assemble for military exercise.

Pistol — The smallest fire-arm used.

Quartermaster — An officer having charge of supplies.

Quarter — Mercy shown by a conqueror.

Quarters — Temporary residence of soldiers.

Regiment — Ten companies of soldiers.

Rampart — An elevation of earth thick enough to resist cannon shot.

Rendezvous — Place of meeting for troops and awaiting orders.

Redan — Two parapets with a ditch in front forming an angle facing the enemy.

Redoubt — A general name for nearly every work in field fortifications, particularly for men inclosing a four-sided area.

Rally — To come back into order.

Re-entering Angle — One pointing inward in a fortification.

Salient Angle — An angle of a wall pointing outward.

Sortie, or Sally — The issuing of a besieged body of troops to attack the besiegers.

Smoke Ball — A mass of composition used in creating a thick smoke.

Staff — A company of officers attached to different departments.

Sap — A trench.

Subsidiary Troops — Those hired by one nation from another.

Tier of Guns — Those on one deck.

Tent — A roll of linen or lint in surgery; also cloth lodgings for soldiers.

Troops — Soldiers in general.

Troop — A small company of horsemen.

Trooper — A horse soldier.

Trophy — Anything taken in victory.

Trumpet — Musical instrument.

Truce — A suspension of operations by agreement of commanders.

Truncheon — A military staff of command.

Trunnion — The projecting points on a cannon used for its support.

Tube — A tin instrument used in quick firing.

Tumbrel — A cart with two wheels used for carrying tools of the pioneers.

Uniform — Dress like that of others of the same classification.

Van — The front of an army.

Vallum — A ditch or wall.

Voltiguer — A light horseman or dragoon.
"In the United States Service each dragoon or horseman has a foot soldier attached to him, who, in case of

necessity, mounts behind on the same horse; thus presenting, whenever they meet the enemy, a kind of infantry and of dragoons in the same regiment."

VOLLEY — A flight of shot. The discharge of many small arms at once.

VOLUNTEERS — One who enters service of his own will.

WEAPON — An instrument of defence.

WAR — A contest between nations or states.

WAY — Passage covered from the enemy's fire.

WARD — A fortress.

WAD — Small bundle of soft material used for keeping the powder and shot close together in a gun.

WATCHWORD — A word given to those who have occasion to pass the guard, so that he may distinguish them from the enemy.

ZOUAVE — The name of a body of soldiers, formerly wearing Arab dress, and introduced into the French army; more recently introduced into the United States service.

A new drill is one of their peculiarities, distinguished from other drills by its seeming irregularities, such as rallying by fours, loading by the soldier in a prostrate condition, &c.

UNITED STATES FORTS.

Fort Monroe.—This fort is a large one, situated near Hampton Roads, Virginia. It has cost over two million dollars. It will accommodate a large garrison within its barracks, and there is space enough inclosed by its walls for an encampment of upwards of five thousand men. The walls are built very thick and strong, and are nearly a mile in length. It is situated a mile and a half from the main land, and completely commands Hampton Roads and the entrance to James River.

It is now garrisoned by 1,100 men.

Fort Riley.—A military post of Kansas, established in 1853, at the junction of Republican and Smoky Hill Forks of Kansas River, on the emigrant route to New Mexico and California, 140 miles from Fort Leavenworth.

It is connected with the latter place by an excellent military road completed to this point in 1854. The Fort has accommodations for a large force of cavalry and stone barracks for eight infantry companies, and, being situated in the midst of a fertile country, abounding in timber, forage, and water, has all the advantages requisite for an important frontier post.

Fort Wayne — Indiana, was abandoned in 1819.

Fort Madison — Iowa, was abandoned in 1813.

Fort Des Moines — Iowa, was abandoned in 1846.

Fort Leavenworth — A military post of Kansas, on the west bank of the Missouri River, 398 miles above its mouth and 31 miles above its junction with the Kansas River.

It is important as a general rendezvous for troops for the West, and as a depot for all the forts on the great Santa Fé and Oregon routes. It is the intersecting point of nearly all the great military roads of the territories, one running south into Texas, one running southwest to Santa Fé, one west to Fort Riley, and a fourth northwest to posts in Nebraska, Utah, Oregon, California, &c.

It is rapidly improving in appearance, being laid out in streets on which stand buildings for the troops, warehouses, quartermaster's establishment, stalls for 8,000 horses, 15,000 mules, &c. The barrack is a large edifice three stories high.

The Hospital was built at a cost of $13,000. Connected with the Fort are several large farms. It is two miles from Leavenworth city.

THE HOME FLEET.

	Off. & Men.	Guns.	Tons.
Flag ship steam frigate Minnesota,	400	26	3,200
Steam frigate Powhattan,	50	10	2,415
Steam corvette Brooklyn,	325	14	2,070
Sailing frigate Sabine,	500	50	1,726
Sailing sloop St. Louis,	300	20	700
Cumberland,	300	24	1,726
Steamer Water-Witch,	62	2	378
Steamer Crusader,	110	10	549
Steamer Mohawk,	110	8	464
Steamer Wyandotte,	100	4	380
Steamer Pocahontas,	95	4	320
Steam gunboat Pawnee,	100	4	1,289

The Macedonian is not included because she is in the gulf of Mexico, and the Falmouth and Supply because they are mere transport storeships.

List of the United States Forts, with the Cost of their Construction.

Fort Wayne, near Detroit, Michigan,	$171,755
Fort Porter, near Buffalo, New York,	116,500
Fort Niagara, Niagara River, New York,	68,027
Fort Ontario, Oswego, New York,	78,013
Fort Montgomery, at outlet of Lake Champlain,	295,960
Fort Knox, narrows of Penobscot River, Me.	381,422
Fort at entrance to Kennebec River, Me.	1,000
Fort Preble, Portland Harbor, Me.	52,350
Fort Scammell, Portland Harbor,	59,840
Fort on Hog Island Ledge, Portland Harbor,	84,200
Fort McCleary, Portsmouth Harbor, N. H.	20,590
Fort Constitution, Portsmouth Harbor, N. H.	17,691
Fort Independence, Boston Harbor,	519,680
Fort Winthrop, Boston Harbor,	150,000
Fort Warren, Boston Harbor,	1,208,000
Fort at New Bedford Harbor, Mass.	95,000
Fort Adams, Newport Harbor, R. I.	1,691,350
Fort Trumbull, New London Harbor, Ct.	250,900
Fort Griswold, New London Harbor, Ct.	14,740
Fort Schuyler, East River, New York,	903,000
Fort opposite Schuyler, East River, New York,	131,000
Fort Columbus, Castle William, & S. Batt., N. Y.	304,460

Fort Gilson, Ellis Island, New York,	$5,100
Fort Wood, Bedloe's Island, New York,	223,000
Fort Richmond, Staten Island, New York,	620,780
Fort on site of Fort Tompkins, Staten Island,	115,800
Battery Hudson, Staten Island,	25,090
Battery Morton, Staten Island,	3,500
Fort Lafayette, at the Narrows, New York,	351,940
Fort Hamilton, at the Narrows, New York,	635,750
Fort at Sandy Hook, New York,	100,000
Fort Mifflin, near Philadelphia, Pa.	83,000

The following are south of "Mason and Dixon's line."

Fort Delaware, Pea Patch Island, Del. River,	339,920
Fort McHenry, Baltimore Harbor, Md.	146,670
Fort Carrol, Saller's Point flats, Balt. Harbor,	703,370
Fort Madison, Annapolis Harbor, Md.	45,600
Fort Severn, " "	6,480
Fort Washington, on Potomac River,	575 370
Fort Monroe, Hampton Roads, Va.	2,476,770
Fort Calhoun, " "	1,824,850
Fort Macon, Beaufort Harbor, N. C.	463,790
Fort at mouth of Cape Fear River,	571,220
Castle Pinckney, Charleston Harbor, S. C.	53,800
Fort Moultrie, " "	87,600
Fort Sumter, " "	977,400
Fort Pulaski, mouth of Savannah River,	988,000

Fort Jackson, Savannah River,	$182,000
Fort Clinch, Amelia Island, Fla.	170,000
Fort Marion, St. Augustine, Fla.	51,400
Fort Taylor, Key West, Fla.	1,130,000
Fort Jefferson, Garden Key, Fla.	1,122,000
Fort Pickens, Pensacola Harbor,	774,000
Fort McRea, Pensacola Harbor,	444,420
Barrancas Barracks and Redoubt, Pen. Har.	598,504
Fort Morgan, Mobile Point, Ala.	1,242,550
Fort Gaines, Dauphin Island, Mobile Bay,	221,500
Defences for Inner Passes, " "	150
Fort on Ship Island, Coast of Miss.	30,190
Military Defence at Proctor's Landing, La.	150,000
Lower Dupree, Bayou Dupree, La.	38,980
Battery Bienvenue, Bayou Bienvenue, La.	129,570
Fort Macomb, Chef Monteur Pass, La.	465,990
Fort Pike, Rigold's Pass, La.	473,000
Fort Jackson, Mississippi River, La.	830,609
Fort St. Philip, opposite Fort Jackson, La.	258,734
Fort Livingston, Barataria Bay, La.	362,370
Defences for Galveston Harbor, Texas,	500
Fort at Fort Point, San Francisco Bay, Cal.	1,553,834
Fort at Alcalnez Island, near San Francisco,	896,660

www.ingramcontent.com/pod-product-compliance
Lightning Source LLC
LaVergne TN
LVHW011146110826
845150LV00008B/2542